Nature's RECORD-BREAKERS

Seas and Oceans

Nature's Record-Breakers

Seas and Oceans

Written by Antonella Meucci

Illustrated by Matteo Chesi, Lorenzo Cecchi, Davide Di Taranto, Fiammetta Dogi, Paolo Holguín, and Antonella Pastorelli

Gareth Stevens Publishing
A WORLD ALMANAC EDUCATION GROUP COMPANY

For a free color catalog describing Gareth Stevens' list of high-quality books
and multimedia programs, call 1-800-542-2595 (USA) or 1-800-461-9120 (Canada).
Gareth Stevens Publishing's Fax: (414) 332-3567.

Gareth Stevens Publishing would like to thank Jan Rafert of the Milwaukee County Zoo,
Milwaukee, Wisconsin, for his kind and professional help with the information in this book.

Library of Congress Cataloging-in-Publication Data

Meucci, Antonella.
    Seas and oceans / by Antonella Meucci ; illustrated by Matteo Chesi ... [et al.].
       p. cm. -- (Nature's record-breakers)
    Includes bibliographical references and index.
    Summary: Provides an assortment of facts about life in the sea, including what science
deems the first multicellular organisms, the world's highest mountain, and the longest alga.
     ISBN 0-8368-2475-X (lib. bdg.)
    1. Marine biology--Juvenile literature. 2. Marine ecology--Juvenile literature. [1. Marine
animals--Miscellanea. 2. Marine biology--Miscellanea. 3. Marine ecology--Miscellanea.
4. Ecology--Miscellanea.] I. Chesi, Matteo, ill. II. Title. III. Series.
QH91.16 .M53   2000
578.77--dc21   00-023897

This edition first published in 2000 by
Gareth Stevens Publishing
A World Almanac Education Group Company
330 West Olive Street, Suite 100
Milwaukee, Wisconsin 53212 USA

Original edition © 1999 by McRae Books Srl. First published in 1999 as *Seas & Oceans*,
with the series title *Blockbusters!*, by McRae Books Srl., via de' Rustici 5, Florence, Italy.
This edition © 2000 by Gareth Stevens, Inc. Additional end matter © 2000 by Gareth Stevens, Inc.

Translated from Italian by Phil Goddard, in association with First Edition Translations, Cambridge
Designer: Marco Nardi
Layout: Ornello Fassio and Adriano Nardi
Gareth Stevens editors: Monica Rausch and Amy Bauman
Gareth Stevens designer: Joel Bucaro

All rights reserved. No part of this book may be reproduced, stored in a retrieval system, or
transmitted in any form or by any means electronic, mechanical, photocopying, recording,
or otherwise without the prior written permission of the copyright holder.

Printed in the United States of America

1 2 3 4 5 6 7 8 9 04 03 02 01 00

Contents

Early Life 6
Seas and Oceans 8
Fish 10
Marine Mammals 12
Other Marine Animals 14
Plants 16
Coral Reefs 18
Polar Seas 20
Ocean Levels 22
Natural Phenomena 24
Exploration 26
More Records 28
Glossary 30
Books, Videos, Web Sites 31
Index 32

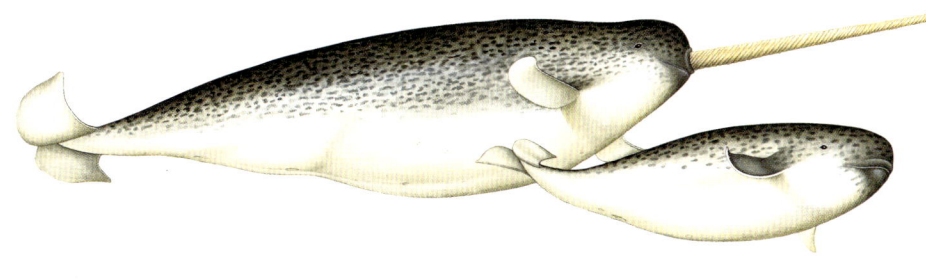

Words that appear in the glossary are printed in **boldface** type the first time they occur in the text.

Did you know?

Q. HAS THERE ALWAYS BEEN LIFE ON EARTH?

A. No. When Earth formed 4.5 billion years ago, its **environment** was harsh and lifeless.

Q. HOW DID LIFE BEGIN?

A. Long processes created **molecules** that were able to multiply, or reproduce. As the molecules grew more complex, they evolved into the first single-celled organisms.

Q. WHAT WERE THE FIRST ORGANISMS?

A. Bacteria and blue-green algae were the first organisms. They appeared in oceans 3.5 billion years ago.

Q. WHY WERE BLUE-GREEN ALGAE SO IMPORTANT?

A. Blue-green algae produced oxygen, which helped other, more complex life-forms develop. Over millions of years, some animals left the ocean and began living on land.

Early Life

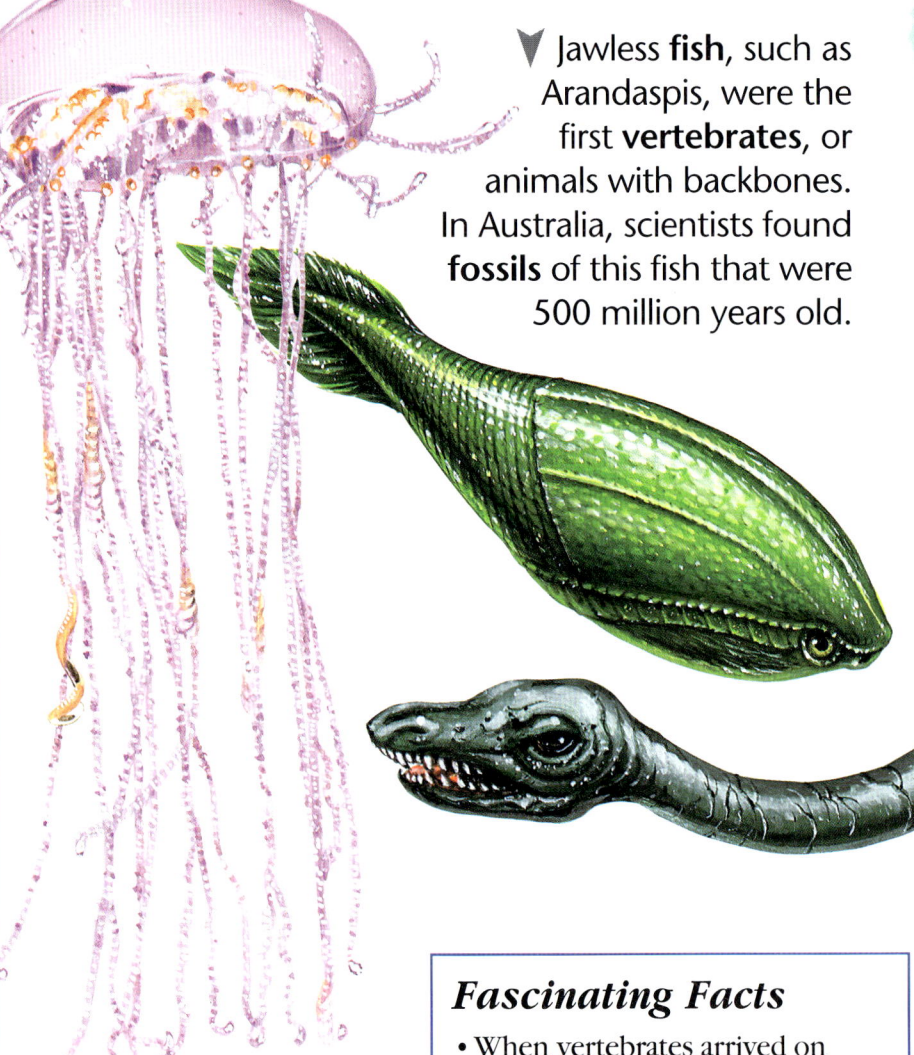

▼ Jawless **fish**, such as Arandaspis, were the first **vertebrates**, or animals with backbones. In Australia, scientists found **fossils** of this fish that were 500 million years old.

▲ Jellyfish and sponges were the first **multi-cellular** organisms, made of more than just one cell. Jellyfish first appeared over 700 million years ago.

Fascinating Facts

• When vertebrates arrived on land, plants and **invertebrates** were already there. Some of these invertebrates were huge. Arthropleura, for example, was a millipede almost 7 feet (2 meters) long.

• Living organisms change into new species through a long process called **evolution**.

◀ Eusthenopteron was the first fish to leave the water. This fish lived about 350 million years ago. It was able to drag itself out of the water and "walk" through the mud using its powerful fins.

Fascinating Fact
Fossils are the remains of prehistoric organisms that have been buried under layers of rock for millions of years. We can study fossils to learn about plants and animals that are now **extinct**.

◀ Elasmosaurus, a plesiosaur, was the longest marine reptile. Plesiosaurs grew to about 46 feet (14 m) long — and half of this was neck! They were good hunters.

▲ Trilobites were the most common ancient animals and the first animals to develop eyes. Some ten thousand species of trilobites lived in the oceans between 540 and 245 million years ago.

Seas and Oceans

▼ An extinct volcano in the Hawaiian Islands is the highest mountain in the world. Including the part that is underwater, this mountain is over 29,529 feet (9,000 m) high. That makes it taller than Mount Everest.

▼ The abyssal plains of the oceans are the greatest plains. These huge, flat areas stretch for hundreds of miles (kilometers) beneath the oceans.

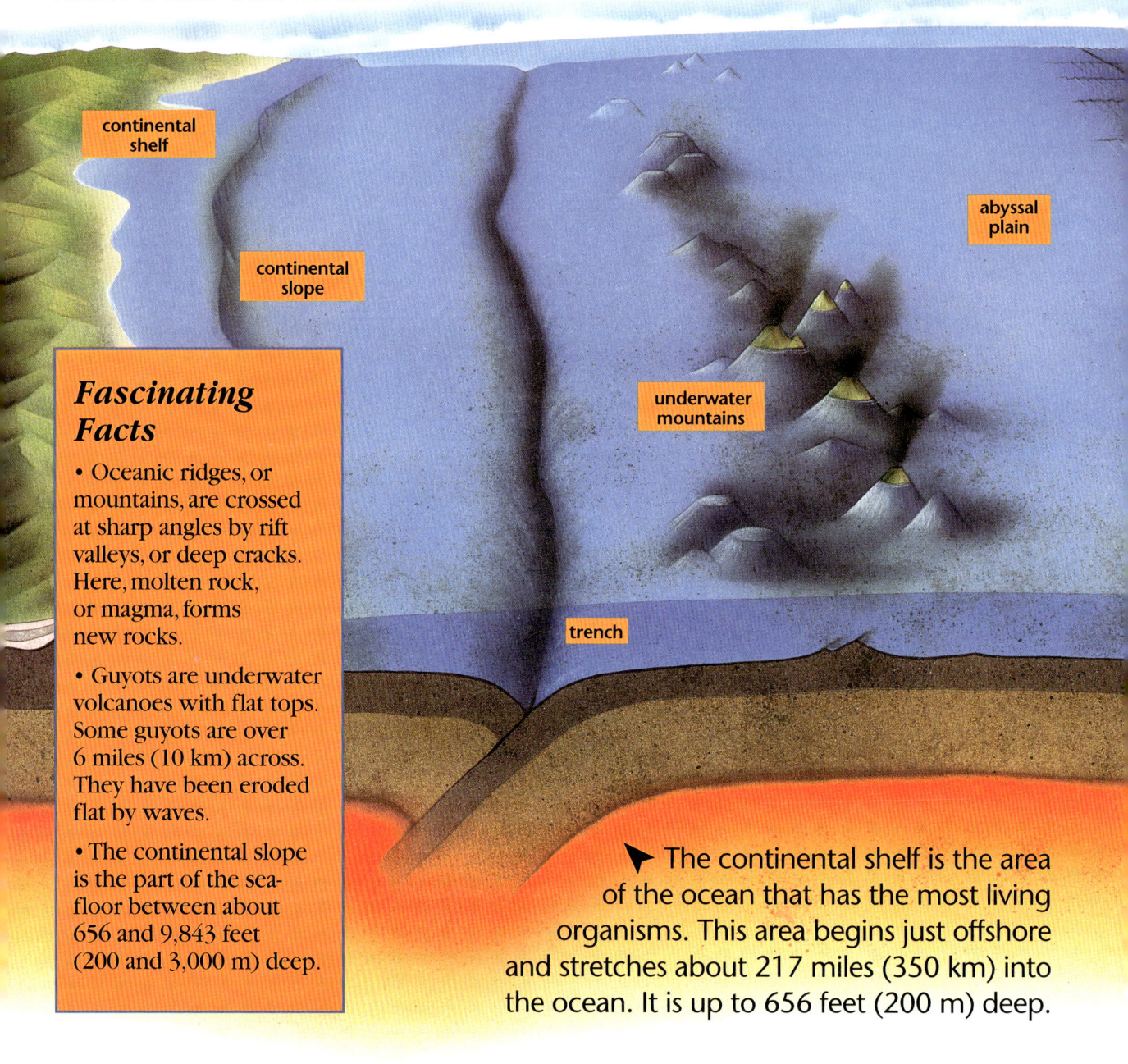

continental shelf

continental slope

abyssal plain

underwater mountains

trench

Fascinating Facts

• Oceanic ridges, or mountains, are crossed at sharp angles by rift valleys, or deep cracks. Here, molten rock, or magma, forms new rocks.

• Guyots are underwater volcanoes with flat tops. Some guyots are over 6 miles (10 km) across. They have been eroded flat by waves.

• The continental slope is the part of the sea-floor between about 656 and 9,843 feet (200 and 3,000 m) deep.

▼ The continental shelf is the area of the ocean that has the most living organisms. This area begins just offshore and stretches about 217 miles (350 km) into the ocean. It is up to 656 feet (200 m) deep.

The Pacific Ocean is the largest, deepest ocean. It has an area of 69 million square miles (179 million square km). That is half the area of all the other oceans! Its average depth is 14,043 feet (4,280 m).

▶ The world's underwater mountains, or oceanic ridges, form the longest mountain chain on Earth. This chain extends 39,770 miles (64,000 km) in length.

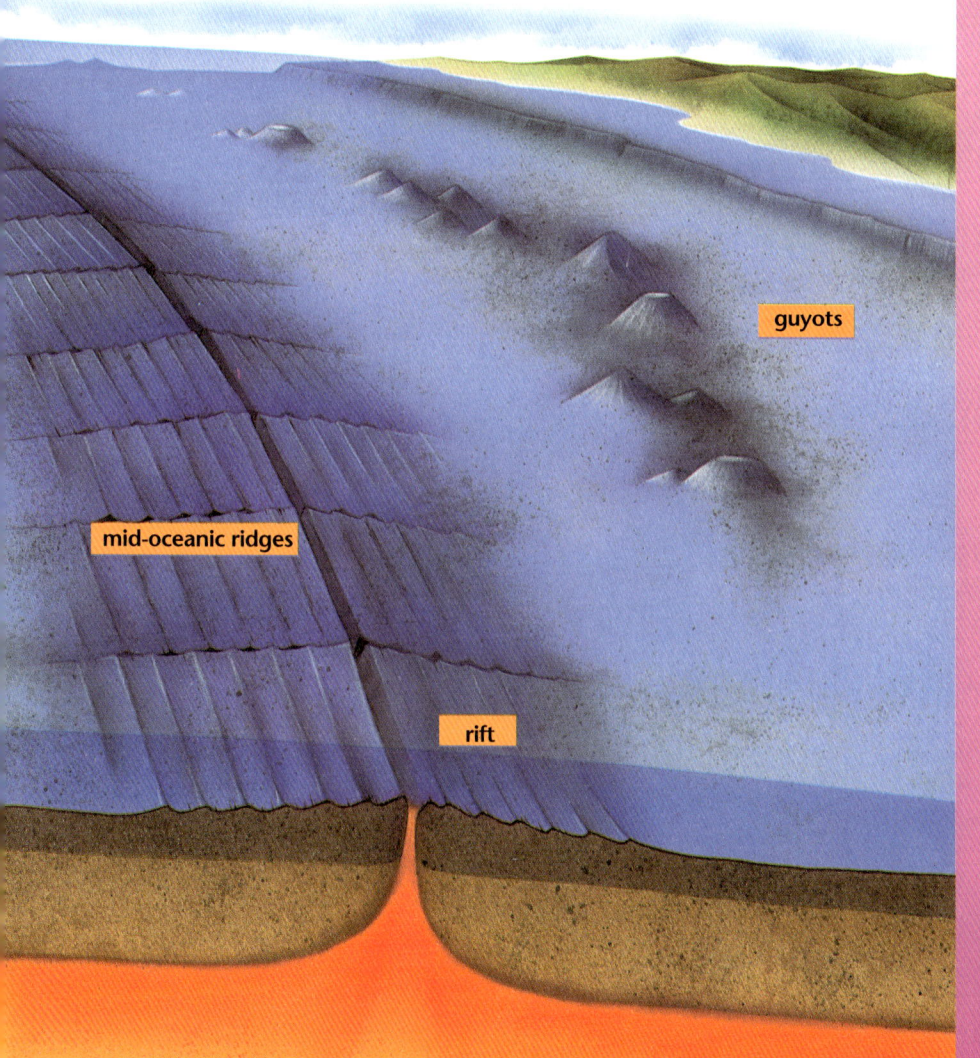

guyots

mid-oceanic ridges

rift

▶ The Mariana Trench in the Pacific Ocean is the deepest trench. Trenches are long, narrow depressions on the ocean bottom. The Mariana Trench is 36,203 feet (11,034 m) deep.

Did you know?

Q. How big are the oceans?

A. The oceans cover about 139 million square miles (360 million sq km) of Earth. This is equal to 71 percent of Earth's surface.

Q. How many oceans does Earth have?

A. Earth has five oceans: the Pacific; the Atlantic; the Indian; the Arctic; and the Antarctic, or Southern, oceans.

Q. What are seas?

A. Seas are bodies of salt water that are smaller than oceans. Some seas, such as the Caspian Sea and the Black Sea, are surrounded by land.

Q. What do the bottoms of seas and oceans look like?

A. People once thought ocean and seafloors were flat. We now know these areas have a variety of features, just like dry land.

Did you know?

Q. What is a fish?

A. A fish is a vertebrate, or an animal with a backbone, that lives in water. Most fish are cold-blooded and use gills to breathe.

Q. How many different species of fish exist?

A. Scientists have identified over twenty thousand species of fish, but there are probably more species that have not yet been discovered.

Q. What are the main groups of fish?

A. Fish are divided into three main groups: jawless fish, such as lampreys and hagfishes; cartilaginous fish, with skeletons made of **cartilage**, or gristle, such as sharks and rays; and bony fish, such as groupers and eels. Bony fish are the largest group.

Fish

➤ The Manta ray is the largest ray. It is also called the "sea devil" because of its appearance. Manta rays can have a wingspan of 23 feet (7 m) and weigh over 3,308 pounds (1,500 kilograms). They eat small fish and **plankton**.

◄ Salmon have the most difficult route to migrate. To lay their eggs, salmon travel hundreds of miles (km) to reach the rivers where they were born. They must battle their way against strong currents and sometimes up waterfalls. After laying their eggs, many salmon die of exhaustion.

▶ The sailfish is the fastest fish. The sailfish lives in tropical waters and can swim at over 62 miles (100 km) an hour.

◀ The cod is the most fertile fish. A single female Atlantic cod can lay up to nine million eggs a season. The eggs float in the water until the fry, or young fish, hatch. Only a few eggs will grow into adults, however, as other fish eat many of the eggs and fry.

◀ The sea horse is a uniquely shaped fish. It moves in an upright position and is a poor swimmer, but its tail is helpful. The sea horse uses its tail to cling to algae and coral. The male sea horse has another unique feature — a pocket on its stomach where the female lays her eggs.

Fascinating Facts

• The lamprey is a jawless fish. It uses its round mouth as a suction cup to cling to other fish. With its sharp teeth, this **parasite** then gnaws the fish's flesh and sucks its blood.

• Some fish can create an electric current strong enough to stun a human. The electric ray, for example, can produce a current of up to 200 volts to protect itself or kill prey.

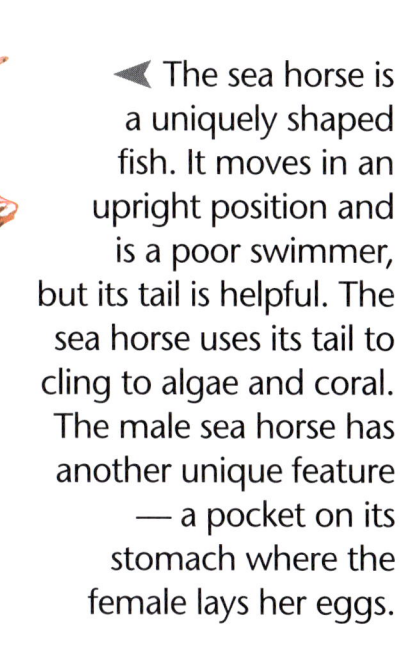

Marine Mammals

◀ The manatee is the gentlest marine **mammal**. It is a distant relative of the elephant and is not at all aggressive. Manatees are plant-eaters, or herbivores, and eat up to 66 pounds (30 kg) of seaweed a day.

▶ The sea otter is the smallest marine mammal. The otter floats on its back when it eats and uses its stomach as a table. When it sleeps, the sea otter wraps itself in seaweed so it won't float away.

▶ The humpback whale has the longest fins. Its pectoral fins, or flippers, can measure up to 16 feet (5 m) long. These huge whales are famous for their "singing." They communicate with each other using complex sounds.

➤ The killer whale is the marine mammal with the biggest appetite. It eats large animals such as seals, dolphins, and even other whales.

Fascinating Facts

• The world's largest animal is the blue whale. It reaches lengths of 89 feet (27 m). Its young can be 20 feet (6 m) long at birth.

• Not all dolphins live in the sea. Freshwater dolphins swim in the rivers of South America and Asia.

• Mammals nurse their young. Seal milk is very rich in fats and helps seal pups grow quickly.

• Beaked whales live in the deep waters of all the world's oceans. They have only one or two pairs of teeth and a long, pointed snout.

◄ The Weddell seal dives longer and deeper than any other seal. It can remain underwater for 43 minutes and reach depths of 1,969 feet (600 m).

Did you **know?**

Q. WHAT IS A MARINE MAMMAL?

A. A marine mammal is a mammal that is fully adapted to living in the sea. Some marine mammals cannot live on land.

Q. WHAT IS THE MAIN DIFFERENCE BETWEEN A MARINE MAMMAL AND A FISH?

A. Marine mammals are warm-blooded animals that use lungs to breathe. They get oxygen from the air, so they must come to the surface to breathe. Fish, on the other hand, use their gills to breathe oxygen from the water.

Q. WHAT KINDS OF MAMMALS LIVE IN THE SEA?

A. Mammals living in the sea include **cetaceans**, such as whales and dolphins; **pinnipeds**, such as seals, otters, and walruses; and **sirenians**, such as dugongs and manatees.

Other

Did you know?

Q. IN ADDITION TO FISH AND MARINE MAMMALS, WHAT OTHER KINDS OF ANIMALS LIVE IN THE SEA?

A. Reptiles, birds, and invertebrates also live in the sea or depend on the sea for survival.

Q. WHAT KINDS OF INVERTEBRATES LIVE IN THE SEA?

A. Invertebrates that live in the sea include squid, octopuses, jellyfish, and starfish. Invertebrates are animals that do not have backbones.

Q. WHERE DO SEABIRDS LIVE?

A. Some seabirds live near coastal waters or on islands. Others live on the open sea. Some seabirds can fly long distances without setting foot on land.

Q. WHERE DO SEA TURTLES LIVE?

A. Sea turtles live mainly in tropical waters. Their big, flat flippers make them agile in the sea but clumsy on land.

◀ The sea snake has the most powerful poison. Sea snakes live in the Indian and Pacific oceans and can grow to 10 feet (3 m) in length. They use their venom to kill prey.

▶ The pearly nautilus is the oldest known **cephalopod**, a scientific **class** that includes squid and octopuses. It is the only cephalopod to have an outer shell, just like its ancestors did 180 million years ago. The nautilus builds its spiral-shaped shell in chambers as it grows.

14

Marine Animals

▼ The wandering albatross has the largest wingspan of any bird. Its wingspan reaches 10 feet (3 m). It flies long distances in search of food, and lands on dry ground only to mate.

Fascinating Facts

• The male frigate bird has a bright red sac under its throat, which it can inflate to the size of a person's head. It inflates this sac during the mating season to attract females.

• Brittle stars, or sand stars, live on the seafloor. They are similar to starfish but have longer, thinner arms.

• Hermit crabs have soft bodies. To protect themselves from **predators**, they crawl into the empty shells of other sea creatures.

▲ The marine iguana is the only lizard to swim in the sea. With its powerful tail, it moves easily through the water, looking for algae to eat. It lives in the Galápagos Islands, where it likes to lie in the sun.

▶ Scallops have the most highly developed eyes of any **bivalve**. They have up to forty eyes along the edges of their shells. They can distinguish only light from dark, but this is enough to help them escape predators.

Plants

Did you know?

Q. Besides animals, what other organisms live in the sea?

A. Plants and algae also live in the sea. They need sunlight to survive, so plants and algae do not exist at depths of more than about 656 feet (200 m).

Q. What are algae?

A. Algae are plantlike organisms. Photosynthesis occurs within their cells, but, unlike plants, algae have no roots, leaves, flowers, or fruit. Some algae are small and single-celled. Other algae, called seaweed, are multicellular. Seaweed can have leaflike fronds and rootlike parts, called holdfasts, that anchor it to the ocean bottom.

Q. What different kinds of algae exist?

A. Algae are grouped by color — green, brown, and red. Most red and green algae are small. Most brown algae grow very large and thrive in cold waters.

◀ Diatoms are the most plentiful species of **microscopic**, single-celled algae. Scientists have identified over fifty thousand different kinds. A diatom is different from other algae because its cell is enclosed in a shell.

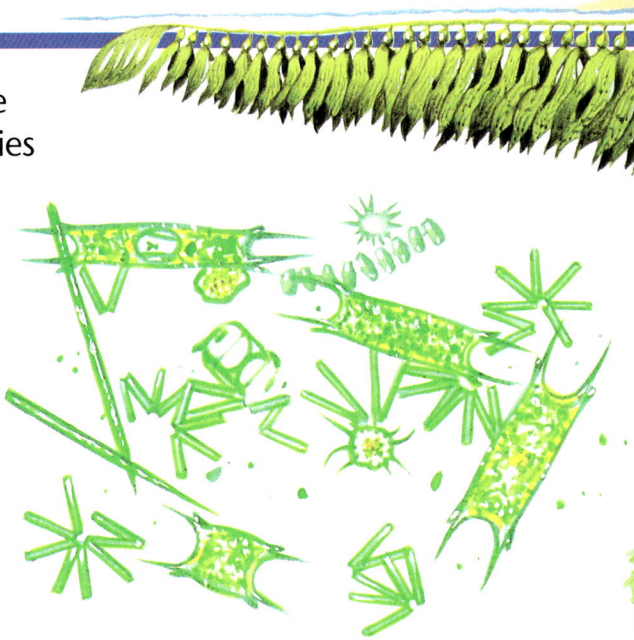

▶ Mangroves are the only trees that grow in seawater. Some mangroves have tangled roots that grow out of the water. Mangrove forests grow along tropical coasts and are home to many animals.

◄ Giant kelp are the longest algae. These brown algae grow in the Pacific Ocean and reach lengths of 164 to 197 feet (50 to 60 m).

Fascinating Facts

• Green plants use the chlorophyll in their leaves to change energy from the Sun into food. This process is called photosynthesis.

• Algae produce oxygen. About 70 percent of the oxygen in the atmosphere is created by algae during photosynthesis.

• In Japan and other parts of the world, seaweed is a delicacy. It is rich in nutrients, such as minerals, proteins, and vitamins.

• Seaweed is used in a variety of products, from paints to lipsticks.

► *Posidonia* forms huge underwater grasslands. This sea grass provides food and shelter for a variety of animals. It produces flowers and huge, olivelike fruit.

◄ *Acetabularia* is one of the smallest green algae. This umbrella-shaped organism lives in shallow water, where there is plenty of sunlight.

Did you know?

Q. HOW DO CORAL REEFS FORM?

A. Coral reefs are built by groups of tiny organisms called polyps. Polyps build a protective shell, or "skeleton." After they die, other polyps build on this skeleton. Reefs form in this way over thousands of years. Polyps live in clear, shallow water in tropical seas.

Q. HOW MANY DIFFERENT TYPES OF CORAL REEFS EXIST?

A. Coral reefs can be divided into three main types: fringing reefs, which lie close to the coast; barrier reefs, found farther from shore; and atolls, circular reefs formed around underwater volcanic islands.

▼ The Great Barrier Reef is the largest coral reef. It lies off the coast of Queensland in Australia. It is nearly 1,243 miles (2,000 km) long and is home to about 400 types of coral and 1,500 species of fish.

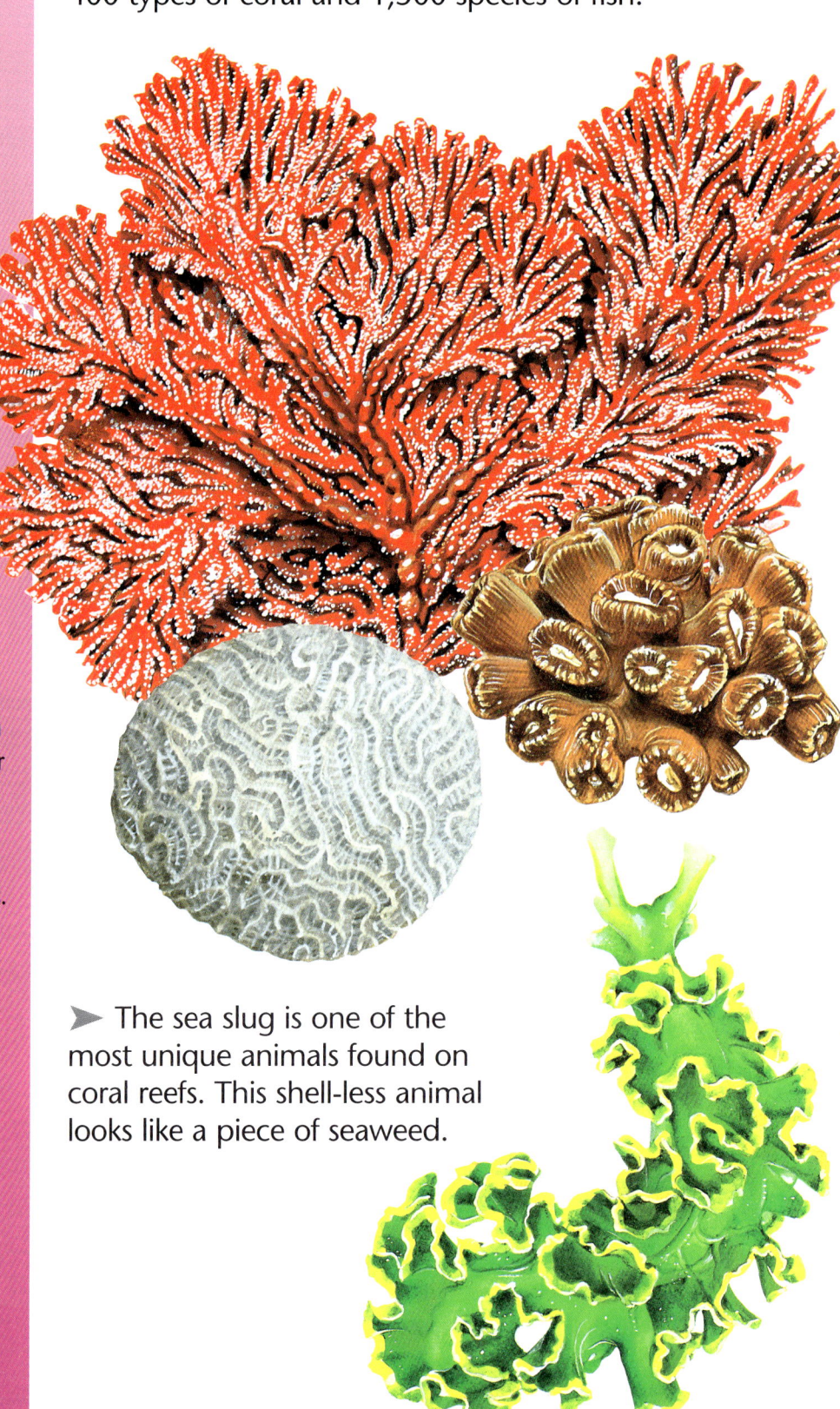

➤ The sea slug is one of the most unique animals found on coral reefs. This shell-less animal looks like a piece of seaweed.

Coral Reefs

◄ The clownfish and the sea anemone are "partners." The clownfish protects itself by living among the anemone's poisonous tentacles, which don't hurt it. In return, the fish cleans the anemone.

▼ Humans do the worst damage to coral reefs. Crown-of-thorns starfish can also damage large areas of reef. This animal eats the polyps in the reef, leaving the coral skeletons empty. This allows the sea to begin eroding the reef.

Fascinating Facts

• The parrot fish wraps itself in protective mucus before it goes to sleep. It takes about half an hour to make enough mucus to cover itself.

• Fire corals have polyps with stinging tentacles. They are dangerous. Simply brushing against them can cause painful burns.

• The cleaner shrimp cleans fish by picking off parasites with its pincers and eating them. It is small and brightly colored.

► Sponges are the simplest multi-celled organisms. They are made up of groups of cells. Different cells in the group carry out different functions. For example, some cells gather food, while others digest it.

Polar Seas

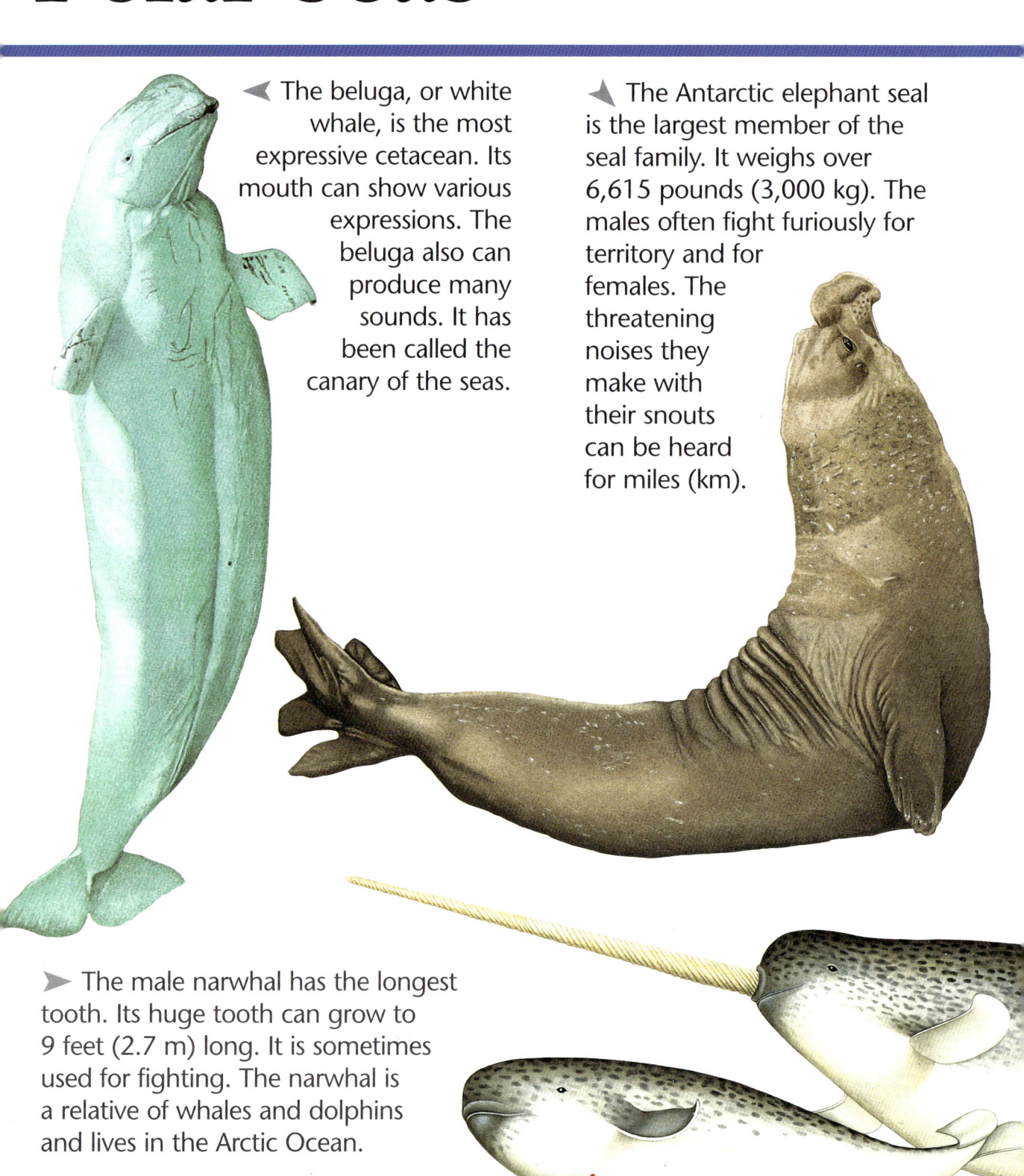

◀ The beluga, or white whale, is the most expressive cetacean. Its mouth can show various expressions. The beluga also can produce many sounds. It has been called the canary of the seas.

◀ The Antarctic elephant seal is the largest member of the seal family. It weighs over 6,615 pounds (3,000 kg). The males often fight furiously for territory and for females. The threatening noises they make with their snouts can be heard for miles (km).

▶ The male narwhal has the longest tooth. Its huge tooth can grow to 9 feet (2.7 m) long. It is sometimes used for fighting. The narwhal is a relative of whales and dolphins and lives in the Arctic Ocean.

20

◣ The emperor penguin is the largest penguin. It lives in Antarctica. In winter, the female lays a single egg. The male looks after the egg, keeping it warm between his feet.

Fascinating Facts

- The walrus is known for its long, white tusks. It uses the tusks for defense and to haul itself out of the water and onto the ice. It lives near the North Pole.

- Penguins cannot fly, but they are strong swimmers. They live only in the cold waters of the Southern Hemisphere.

- Polar bears prey on Arctic seals. The bears will actually wait beside holes in the ice. When seals come up to breathe, the bears kill them with a blow from their powerful paws.

◣ Krill travel in the largest groups, larger than those of any other organisms that live in the sea. Swarms that contain several million of these tiny **crustaceans** can be over 7 miles (12 km) long.

Did you know?

Q. Where is the Arctic Ocean?

A. The Arctic Ocean is located at the North Pole. This ocean and the land surrounding it are mostly covered in ice. This ice does not melt — even during summer.

Q. Where is the Antarctic Ocean?

A. The Antarctic Ocean, also called the Southern Ocean, includes the southern parts of the Atlantic, Pacific, and Indian oceans. These waters surround Antarctica, the frozen continent that has the lowest temperatures on Earth. Few animals survive in Antarctica, but the ocean around it is rich in animal and plant life.

Q. What are icebergs?

A. Icebergs are mountains of ice. They break off of glaciers and drift into the sea. The part of an iceberg below the water is often larger than the part above the water's surface. This makes icebergs a danger to ships.

Ocean Levels

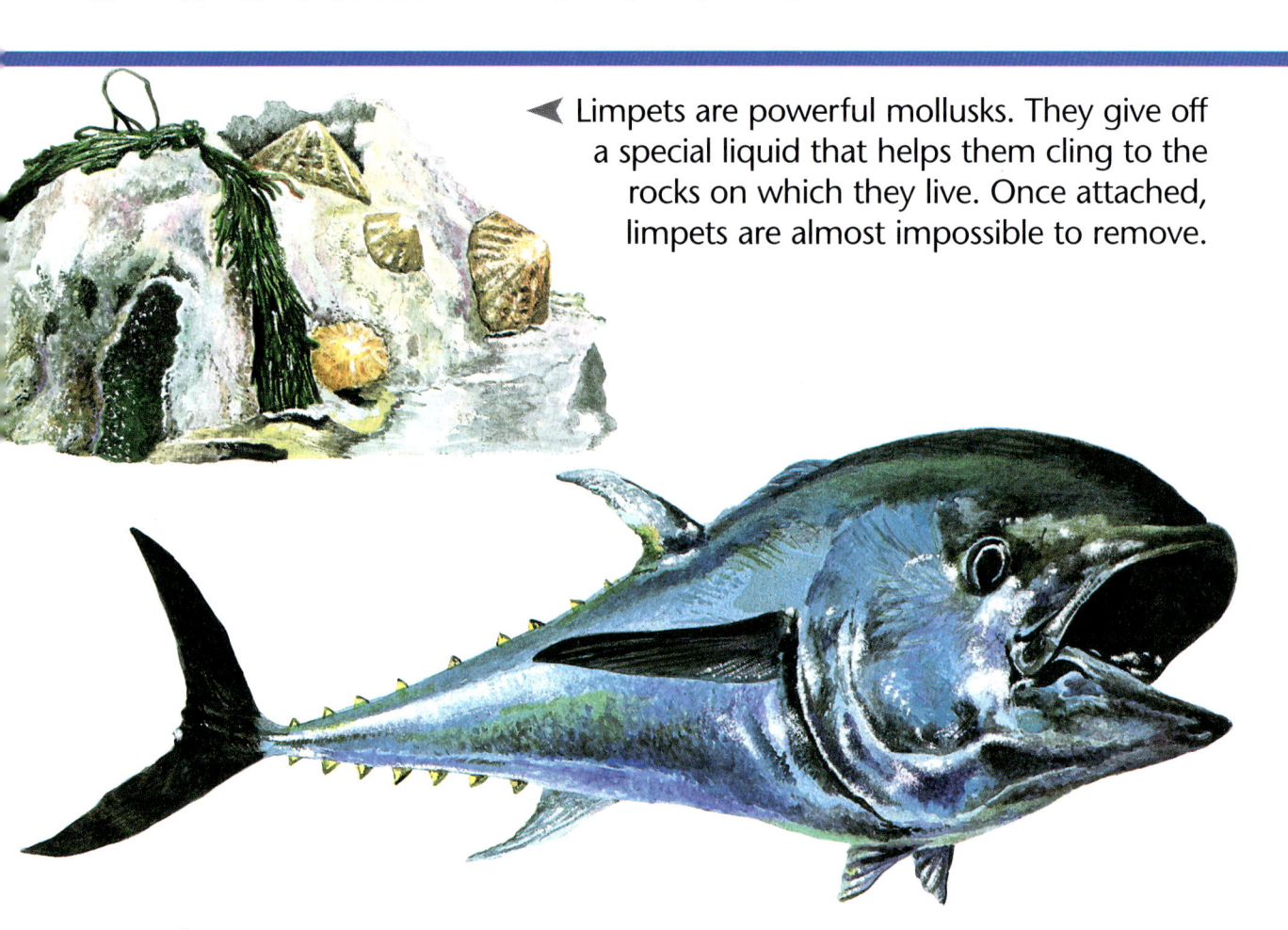

◄ Limpets are powerful mollusks. They give off a special liquid that helps them cling to the rocks on which they live. Once attached, limpets are almost impossible to remove.

◄ The tuna fish is the best and most tireless swimmer. Its shape and features allow it to move through water more easily than most other fish.

◄ The sea spider's body is so tiny that some of its organs, including its reproductive organs, are in its legs. It lives in all the oceans, at many different depths — even at the bottom of the oceanic abysses. The sea spider is bright red and has long, thin legs.

◀ Flying fish are the only fish that can "fly." They have fins that extend like wings. These fins help them skim across the water's surface. They usually fly for twenty to thirty seconds and cover about 328 feet (100 m) at a time.

Fascinating Fact

The deeper the water, the greater the water pressure. Fish living in the deepest waters have body pressures equal to the great water pressure around them. If these fish came to the surface, where the pressure is lower, they would explode!

Did you know?

Q. WHAT ARE THE DIFFERENT LAYERS OF THE OCEAN CALLED?

A. Scientists separate the ocean into layers according to how much sunlight reaches each one. The top layer is the euphotic zone. It receives the most sunlight and reaches down to about 262 feet (80 m). Below this is the disphotic zone, which gets some sunlight and extends to about 656 feet (200 m). The next layer is the aphotic zone, which extends to the ocean floor. This layer receives no sunlight. Each layer has its own animal life.

Q. AT WHAT DEPTH DO MOST SEA CREATURES LIVE?

A. Most sea creatures live in the surface layer, or euphotic zone, which receives the most sunlight. Fewer animals exist with each deeper layer. Still, some animals live at even the greatest depths.

◀ Abyssal fish are fish that live at the ocean's greatest depths. Some have a glowing "lure" on their heads, used to attract prey.

Did you know?

Q. How do waves form?

A. Waves form as wind blows across the surface of water. Seas become rougher as more waves form, or as waves grow.

Q. What are marine currents?

A. Marine currents are huge "rivers" of water that flow through the oceans. Some currents are warm, while others are cold. Some run along the surface of the oceans, while others run below the surface.

Q. Why are marine currents important?

A. Marine currents affect the **climate**. The Gulf Stream, for example, is a warm current that keeps the seas around Scandinavia from freezing in winter.

Q. Can we prevent the destruction caused by hurricanes?

A. No, but new technology is helping us identify and track hurricanes as they form. This means that people can be warned sooner when a hurricane is coming.

▼ The powerful Antarctic Circumpolar Current is the longest marine current on Earth. It is the only current to circle the entire planet.

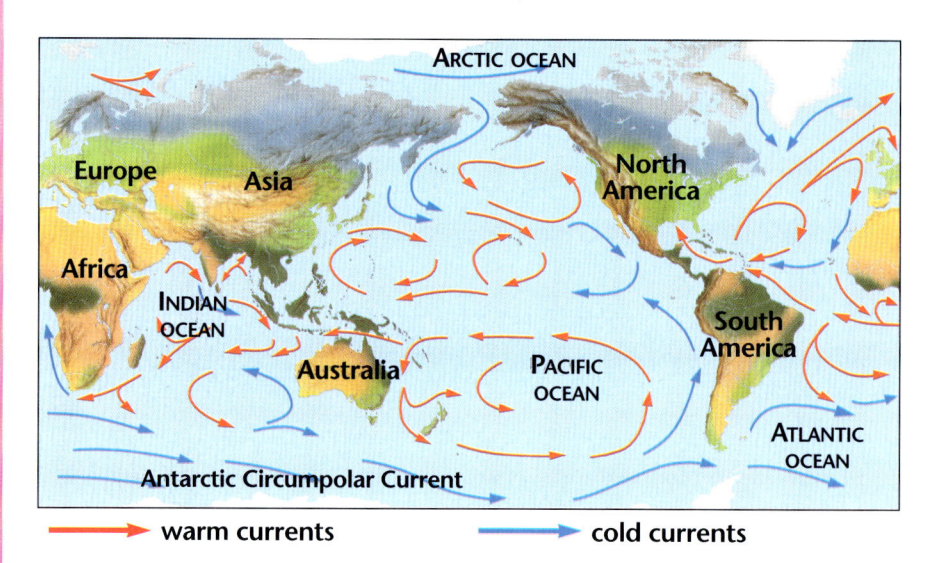

◄ Waterspouts are caused by circular winds traveling up to 50 miles (80 km) an hour. These winds lift water high into the air. When the winds slow down, the water comes tumbling down, sometimes damaging ships.

Natural Phenomena

▶ Hurricanes and cyclones are the most violent storms. They form over warm water and produce very high winds and heavy rain. This satellite photo of a hurricane shows a spiral cloud with the storm's "eye" in the center.

▲ The Bay of Fundy in Canada has the highest **tides**. There, the difference in height between high and low tide is over 49 feet (15 m).

▶ Gigantic waves, called tsunamis, are caused by volcanic eruptions or earthquakes under the sea. These waves can travel across open water at speeds of up to 500 miles (800 km) an hour. When they reach the coast, they may be as high as 295 feet (90 m).

25

Exploration

▶ The British ship *Challenger* made the largest oceanographic **expedition**. It set off in 1872 and traveled across all the oceans except the Arctic, making observations and collecting samples of marine animals and plants.

Fascinating Facts

• Scientists use echo sounding, or sonar, to measure the depth of the sea. They map the seafloor by bouncing sounds off of it and measuring the time it takes those sounds to return to the surface.

• In the nineteenth century, divers wore brass helmets and breathed air through tubes running up to the surface.

◀ In 1977, the Russian icebreaker *Arktika* was the first ship to reach the North Pole. Icebreakers are specially designed to break through thick layers of ice in polar seas.

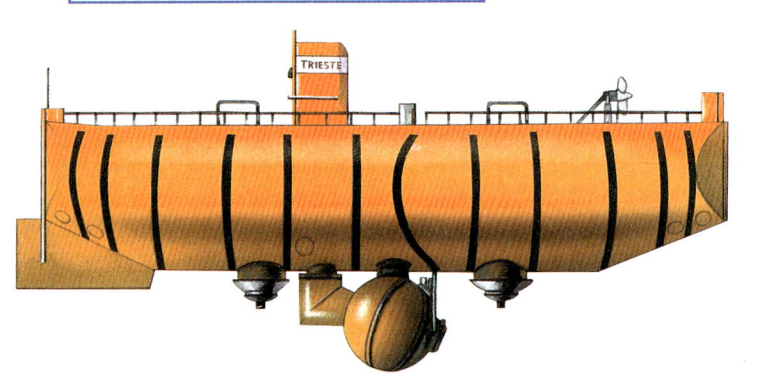

▶ The bathyscaphe *Trieste* completed the deepest dive in 1960. It reached a depth of 35,802 feet (10,912 m) in the Pacific Ocean. A bathyscaphe is designed for underwater exploration.

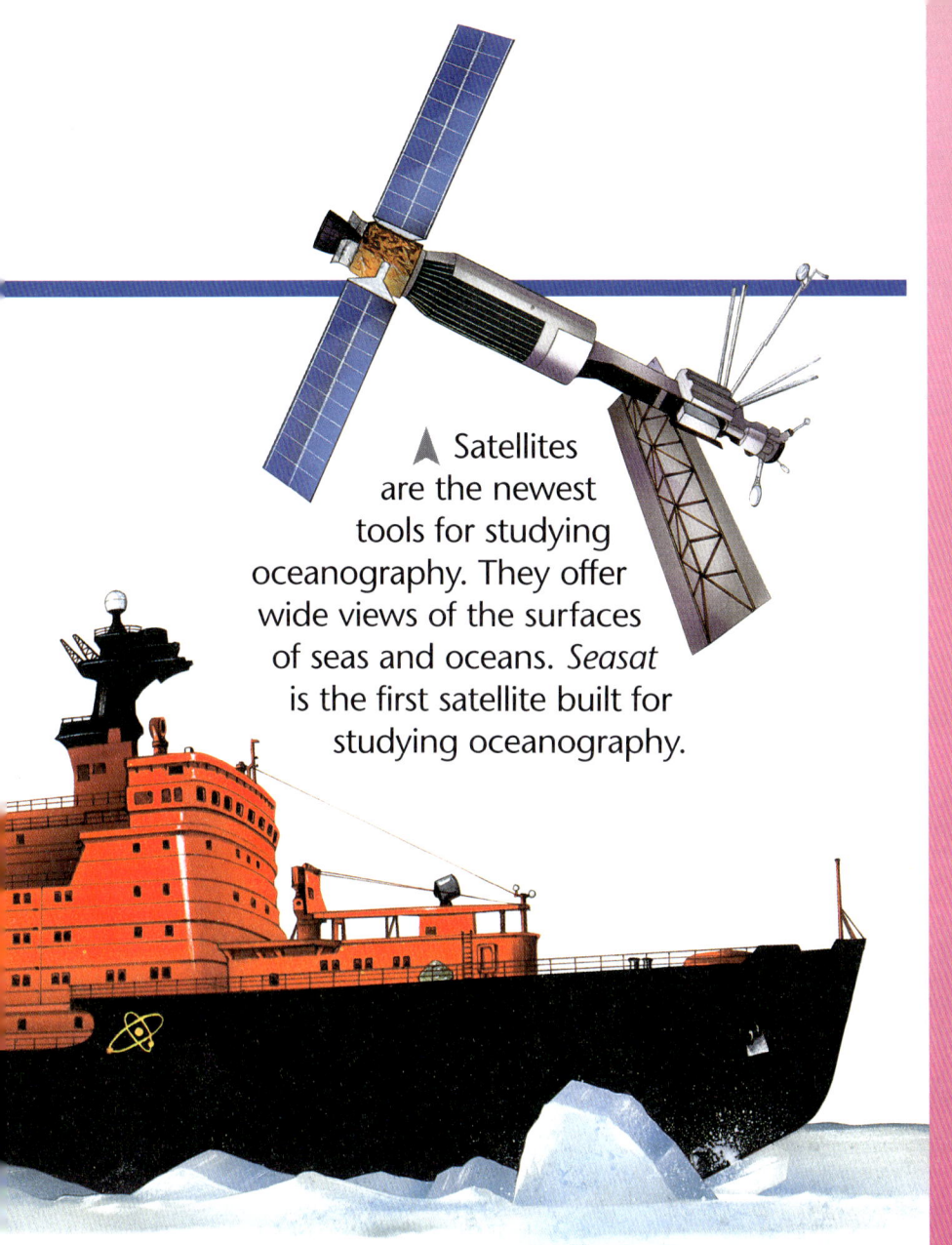

▲ Satellites are the newest tools for studying oceanography. They offer wide views of the surfaces of seas and oceans. *Seasat* is the first satellite built for studying oceanography.

◀ The submarine *Alvin* took the first photographs of chimney-shaped hot-water vents near the Galápagos Islands. These vents are 8,203 feet (2,500 m) below the surface.

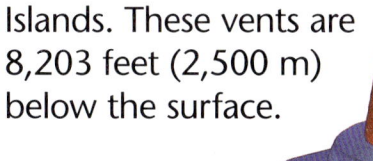

Did you know?

Q. When did people begin to study the sea?

A. Scientists first began studying the sea in the 1700s. Before this, people traveled across the oceans mainly for land exploration.

Q. What is oceanography?

A. Oceanography is the study of seas and oceans. It includes everything from measuring the temperature of the water to studying currents, mapping the seafloor, and studying marine animals and plants.

Q. What is an oceanographic vessel?

A. An oceanographic vessel is a ship or other conveyance that has scientific equipment and laboratories. Scientists can study information gathered from the sea right on board the ship.

27

More Records

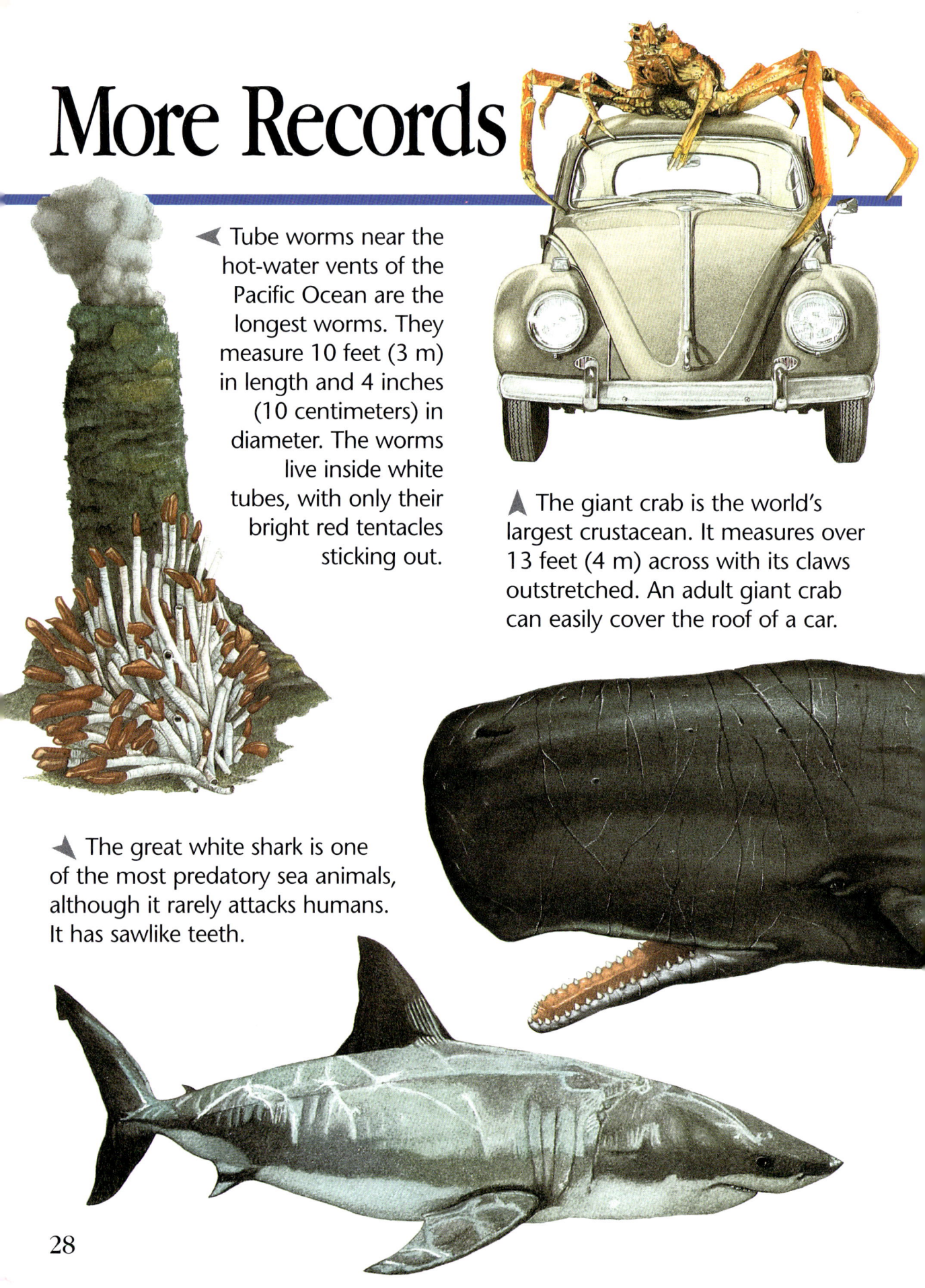

◀ Tube worms near the hot-water vents of the Pacific Ocean are the longest worms. They measure 10 feet (3 m) in length and 4 inches (10 centimeters) in diameter. The worms live inside white tubes, with only their bright red tentacles sticking out.

▲ The giant crab is the world's largest crustacean. It measures over 13 feet (4 m) across with its claws outstretched. An adult giant crab can easily cover the roof of a car.

◀ The great white shark is one of the most predatory sea animals, although it rarely attacks humans. It has sawlike teeth.

▶ The dolphin is the most intelligent sea creature. Dolphins have well-developed learning skills and are social animals. If a dolphin is sick or hurt, its companions will help it float on the surface to keep it from drowning.

▼ The sperm whale has the largest brain of any animal. The sperm whale's head makes up a third of its total body length. This cetacean dives to great depths in search of its favorite food, giant squid.

Fascinating Fact

Echolocation is a technique used by dolphins and other cetaceans to navigate. The animal emits high-frequency sounds that echo back when they hit something. This helps the animal locate obstacles and other creatures without actually seeing them. It is the same principle humans use with echo sounding, or sonar.

Did you know?

Q. WHAT IS A FOOD CHAIN?

A. A food chain is a series of living organisms, each of which is a food source for the next one. Fish, for example, are eaten by seals, which in turn are eaten by polar bears.

Q. WHAT IS AN ECOSYSTEM?

A. An ecosystem is the combination of an environment and all the plants and animals that live in that environment. Many parts of an ecosystem are closely related through food chains.

Q. WHICH IS THE MOST IMPORTANT MARINE ECOSYSTEM?

A. Coral reefs contain more animals and plants than any other marine ecosystem. Reefs also contain a complete food chain, from single-celled algae to large predators.

Glossary

bivalve: an animal with a soft body that is enclosed in two shells, or valves. Clams and scallops are bivalves.

cartilage: an elastic substance. Some fish have skeletons made of cartilage instead of bone.

cephalopod: one of a class of aquatic animals with soft bodies and muscular arms around a head. Octopuses and squid are cephalopods.

cetacean: one of a group of aquatic mammals that has a torpedo-shaped body with one or two openings at the top of its heads, two paddle-shaped limbs, and a flat tail. Whales are cetaceans.

class: a group of organisms that shares certain features. Seals and walruses belong to the class *Mammalia*.

climate: the average weather conditions over a period of time. A desert, for example, receives very little rain and therefore has a dry climate.

crustaceans: animals, such as crabs, that have segmented bodies and a shell.

environment: the surroundings in which plants, animals, and other organisms live.

evolution: the process of changing, or developing, over time from one form to another. All living things must evolve in order to survive in a changing environment, or they will become extinct.

expedition: a journey made for a certain reason, such as for studying the oceans.

extinct: (1) no longer alive, such as when all the animals of one species die out. (2) unable to erupt, as with a volcano.

fish: vertebrates that live in water. Most fish are cold-blooded and use gills to breathe.

fossils: traces or remains of animals or plants from an earlier period of time, often found in rock.

invertebrates: animals without backbones.

mammals: warm-blooded animals that feed their babies on milk produced by the mother. Mammals usually give birth to live young.

microscopic: objects or organisms so tiny that a microscope is needed to see them.

molecule: the smallest and most basic particle into which a substance can be divided and still be the same substance.

multicellular: having more than one cell. A cell is the smallest part of a plant or animal.

parasites: animals that live in or on other animals, called hosts, and depend on the hosts for survival.

pinnipeds: aquatic, meat-eating animals with four flippers as limbs. Seals and walruses are pinnipeds.

plankton: tiny plants and animals that drift in the ocean.

predators: animals that hunt and kill other animals for food.

sirenians: aquatic, plant-eating mammals. Dugongs and manatees are sirenians.

tide: the rising and falling of ocean waters based on the gravitational pull of the Sun and the Moon on Earth.

vertebrates: animals that have backbones.

More Books to Read

Colors of the Sea (series). Eric Ethan and Marie Bearanger (Gareth Stevens)

Fish. Wonderful World of Animals (series). Beatrice MacLeod (Gareth Stevens)

The Living World. Record Breakers (series). David Lambert (Gareth Stevens)

Manatee Magic for Kids. Animal Magic for Kids (series). Patricia Corrigan (Gareth Stevens)

Ocean Life. Under the Microscope (series). Casey Horton (Gareth Stevens)

Our Wet World: Exploring Earth's Aquatic Ecosystems. Sneed B. Collard, III (Charlesbridge Publishing)

The Science of Water. Living Science (series). Janice Parker (Gareth Stevens)

Seashells, Crabs, and Sea Stars. Young Naturalist Field Guides (series). Christiane Kump Tibbitts (Gareth Stevens)

Secrets of the Animal World (series). Eulalia García, Isidro Sánchez, and Andreu Llamas (Gareth Stevens)

Videos

Amazing Animals Video: Underwater Animals. (Dorling Kindersley)

In the Company of Whales. (Discovery Home Video)

National Geographic's Ocean Drifters. (National Geographic)

Secrets of the Ocean Realm. (PBS Home Video)

Web Sites

WhaleTimes SeaBed
www.whaletimes.org/

PBS Online: Secrets of the Ocean Realm
www.pbs.org/oceanrealm/index.html

Salt Water Animals
mbgnet.mobot.org/salt/animals/index.htm

Ocean Info
oceanlink.island.net/oinfo/oinfo.html

Temperate Oceans
mbgnet.mobot.org/salt/oceans/index.htm

Under the Sea
www.germantown.k12.il.us/html/sea.html

Some web sites stay current longer than others. For further web sites, use your search engines to locate the following keywords: *coral reefs, dolphins, fish, marine mammals, oceanography, oceans, seals, seas, seaweeds, sharks,* and *whales.*

Index

abyssal fish 23
abyssal plains 8
Acetabularia 17
algae 6, 11, 15, 16, 17, 29
Alvin 27
anemones 19
Antarctic Circumpolar Current 24
Antarctic Ocean 9, 21
Anthropleura 6
Arctic Ocean 9, 20, 21, 24, 26
Arktika 26
Atlantic Ocean 9, 11, 21, 24

belugas 20
birds 14, 15
Black Sea 9
brittle stars 15

Caspian Sea 9
cetaceans 13, 20, 29
Challenger 26
cleaner shrimp 19
clownfish 19
continental shelf 8, 9
continental slope 8
coral reefs 18, 19, 29
crown-of-thorns starfish 19
crustaceans 21, 28
currents 24, 27

diatoms 16
dolphins 13, 20, 29

Elasmosaurus 7
electric rays 11

fire corals 19
fish 6, 7, 10, 11, 13, 14, 18, 19, 22, 23, 29

flying fish 23
fossils 6, 7
frigate birds 15

giant crabs 28
giant kelp 17
Great Barrier Reef 18
great white sharks 28
guyots 8, 9

hermit crabs 15
hot-water vents 27, 28
humpback whales 12
hurricanes 24, 25

icebergs 21
Indian Ocean 9, 14, 21, 24

jawless fish 6, 10, 11
jellyfish 6, 14

krill 21

lampreys 10, 11
limpets 22

mammals 12, 13, 14
manatees 12, 13
mangroves 16
Manta rays 10
Mariana Trench 8
marine iguanas 15

narwhals 20

ocean layers 22, 23
oceanic ridges 8, 9
octopuses 14
otters 12, 13

Pacific Ocean 8, 9, 14, 17, 21, 24, 26, 28
parrot fish 19

pearly nautilus 14
penguins 21
plants 6, 7, 12, 16, 17, 21, 26, 27, 29
polar bears 21, 29
polyps 18, 19
Posidonia 17

rays 10, 11

sailfish 11
salmon 10
satellites 25, 27
sea horses 11
sea slugs 18
sea snakes 14
sea spiders 22
sea turtles 14
seals 13, 20, 21, 29
Seasat 27
seaweed 12, 16, 17, 18
sharks 10, 28
sonar 26, 29
sponges 6, 19
squid 14, 29
starfish 14, 15, 19
submarines 27

tides 25
trenches 8
Trieste 26
Trilobites 7
tube worms 28
tuna fish 22

walruses 13, 21
wandering albatross 15
waterspouts 24
waves 24, 25
Weddell seals 13
whales 12, 13, 20, 29